INTRODUCTION

The United States Navy and United States Marine Corps flew thousands of combat missions during World War II and Korea. The vast majority of these were flown under extremely hazardous conditions requiring total concentration, dedication, and often a generous portion of good luck to get back safely.

USN and USMC pilots were among the finest trained aviators and did themselves proud during the rugged combat conditions of WWII and Korea. However, even with their superior training and the finest aircraft, battle-damage and crashes occurred. Operational accidents while flying from carriers in all-weather conditions claimed a large number of aircraft as did damage sustained during combat, and frequently, shot up aircraft trying to make it back to their carrier crashed on the decks. These photos and the stories that go with them are meant as a tribute to the brave pilots and aircrew that made up this part of Naval Aviation history. Not to be overlooked are the aircraft flown by these hardy aviators —Corsairs, Hellcats, Skyraiders and Panthers just to name a few. These sturdy aircraft, frequently brought their pilots and aircrew back to the carriers and divert fields even though the planes themselves were severely shot-up, some to the point of being *so much flying scrap metal*. These shredded airframes were often pushed over the side of carriers or relegated to the boneyards of the airfields after getting their pilots back. Often after getting their feet back on the ground pilots and aircrew were amazed at the amount of punishment their aircraft had taken and remained in the air. Aircraft returned with huge holes from bullets and anti-aircraft fire, sometimes with entire flying surfaces shredded or shot away.

During wartime, it was not uncommon to see mis-matches of paint where replacement flaps, rudders, elevators or sheet metal patches were applied by Navy metalsmiths. There were numerous instances where different shot-up airframes were matched together to make up a single flyable aircraft out of the remains of several wrecks. One of the best known of these aircraft was *BLUE TAIL FLY*, a pieced together F9F Panther aboard PRINCETON during the Koran War, and there were many others.

The following pages present graphic examples of many of these aircraft and tells much of what is known of them. Many have heard the saying... *any landing you walk away from is a good landing* ...and it is doubtful that any surviving pilot would argue the point.

Some forty years have passed since World War II and Korea, the impact of these experiences are as stunning today as they were then... and thanks to the brave and often hazardous efforts of combat cameramen, we can take a trip back into time and see some of the seldom viewed events that make up a significant part of Naval Aviation history.

Jim Sullivan
Charlotte, NC
December, 1985

1943

(Above) White 4, a PBJ-1 (35046) of Marine Air Wing-3 was hit by another taxiing Mitchell bomber. This PBJ-1 carried the Blue Gray over Dark Gray camouflage scheme and was used to train pilots heading for the South Pacific. MCAS Cherry Point, NC, 24 July 1943. (USMC-National Archives)

(Right) F6F-3 Hellcat (04788) of VF-9 went over the port side while landing aboard USS ESSEX. The gunners just below the flight deck are ducking low as White 33 finishes its last flight. 5 April 1943. (USN-National Archives)

(Below) Black 8, a PBM-3 Mariner of VP-208, has sheered off both floats and is listing dangerously in the waters of the British West Indies. Barrels have been attached to the Mariner to help keep it afloat. 5 May 1943. (USN via Dave Lucabaugh)

BENT & BATTERED WINGS

USN/USMC DAMAGED AIRCRAFT 1943-1953

by Jim Sullivan
illustrated by Don Greer

squadron signal publications

A Battle-torn TBM-3 Avenger of VT-17 limps back to BUNKER HILL after Japanese anti-aircraft fire damaged the wings. 19 February 1945.

COPYRIGHT © 1986 SQUADRON/SIGNAL PUBLICATIONS, INC.
1115 CROWLEY DRIVE, CARROLLTON, TEXAS 75011-5010
All rights reserved. No part of this publication may be reproduced, stored in a retrival system or transmitted in any form by any means electrical, mechanical or otherwise, without written permission of the publisher.

ISBN 0-89747-182-2

If you have any photographs of the aircraft, armor, soldiers or ships of any nation, particularly wartime snapshots, why not share them with us and help make Squadron/Signal's books all the more interesting and complete in the future. Any photograph sent to us will be copied and the original returned. The donor will be fully credited for any photos used. Please send them to: Squadron/Signal Publications, Inc., 1115 Crowley Dr., Carrollton, TX 75011-5010.

DEDICATION

Let us always remember the brave fliers who flew to defend our Country in times of war ... and let us never forget those that did not return.

ACKNOWLEDGEMENTS

The author sincerely wishes to thank all of the people who contributed to this project. Special thanks are in order to Dave Lucabaugh and Clay Jansson for their in-depth photographic coverage; to Bill Love for his assistance in putting it all together; and to Bill Halsey for his contribution of copy camera work.

Hal Andrews	Paul McDaniel
Roger Besecker	Ron Picciani-Aircraft Slides
Henry Covington	Harold Reutebuch
Bill Crimmins	W.E. Scarborough
Bill Halsey	Larry Smalley
Dick Heath	Don Spering, A.I.R.
Dick Hill	Larry Webster
Clay Jansson	Hank Weimer
Bob Lawson, The HOOK	Jim Wiedie
Bill Love	National Archives
Dave Lucabaugh	USN/USMC
Peter Mersky	

Joe Michaels/J.E.M. Aircraft Slides

Smithsonian/National Air and Space Museum:
Dana Bell
Susan Ewing
Larry Wison

(Overleaf) F4U Corsairs, a B-24 Liberator, a PBY Catalina and an F6F Hellcat are all part of the boneyard at Bougainville's Torokina Point Airfield. 4 February 1944. (USMC)

(Below) This PB2B-2 (44243) of VH-2, a Boeing-built Catalina assigned to Air Rescue, damaged the starboard wing, lost the wingtip float, and crumpled the aileron. The USS St George (AV-16) stood by for rescue and repairs. 26 February 1945. (USN via Dave Lucabaugh)

(Above) TBF-1 Avenger carrying the White C-4 code of VC-30 is stopped by the barrier aboard USS MONTEREY. The tailwheel was damaged in the rough landing. June 1943. (Hank Weimer)

(Left) Curtiss SO3C-3 (4257) Seamew is hoisted aboard USS DENVER after flipping over while making an open-water landing during 1943. (USN via Dave Lucabaugh)

F4U-1 Corsairs of VF-17 crashed while landing aboard BUNKER HILL. (Below Left) White 26 nosed over when it engaged the barrier. Damage seems to be confined to the prop and starboard wing. (Below) This early birdcage Corsair flipped over on its back and is being recovered. The plexiglass section on the fuselage bottom at the wing flap line was found on early production F4U-1s permitting the pilot to see directly beneath the plane. July 1943. (USN-National Archives)

(Above) F6F-3 (25974) Hellcat of VF-2 went into the port catwalk while landing. The deck crew were put on the starboard wing to add leverage when hauling White 7 back aboard USS CHARGER. 12 August 1943 (USN-National Archives)

(Left) LT Donald Balch of VMF-221 sits on his parachute beside his F4U-1 (02467) Corsair. Balch made it back to his Russell Islands base in No 125 after combat with Jap fighters. This rugged Corsair survived to fight another day. 6 July 1943. (USMC/Vought)

(Below) SBD-5A (09752) Dauntless of Marine Air Wing-3 marked B-37 went off the runway at MCAS Cherry Point, hit an embankment and broke up. 20 August 1943 (USMC-National Archives)

(Above) This SNJ-3C (01920) of CASU-23 (Carrier Air Service Unit-23) nosed over while landing aboard USS CHARGER. The North American built SNJ trainer was tough on inexperienced pilots who occasionally ended up like this. 26 August 1943 (USN via Dave Lucabaugh)

(Below) SDB-5 (09709) Dauntless of VMSB-331 belly landed with little damage other than a bent prop. White 9 was based at MCAS Cherry Point, NC. 23 August 1943. (USN-National Archives)

(Below Right) This TBF-1 Avenger, Black 24 in the Gray over White Atlantic paint scheme, went off the port side of USS CARD's flight deck and into the sea. The crew escaped safely. 26 December 1943. (USN via Dave Lucabaugh)

(Below) This F4U-1 (02576) of VMF-222 is being recovered after a crash landing that left the Corsair on its back in a drainage ditch. The pilot escaped with only minor injuries but *MARINE DREAM* never flew again. Bougainville, 13 December 1943. (USMC)

(Above and Above Right) This F6F-3 (66101) Hellcat of VF-6 (Det) piloted by LT (jg) A W Magee landed aboard USS COWPENS with his belly tank ablaze. The fire was out in 1½ minutes, with Magee out well before that. 24 November 1943. (USN-National Archives)

(Below) PB2Y-3 *Betty B* carries the codes P-70. This Coronado flying boat was beached near Salinas, Brazil. As bulldozers attempt to turn the boat, sailors used their weight to bring the port wing down. 13 October 1943. (USN via Dave Lucabaugh)

(Below) F6F-3 Hellcat, White C15 has been chewed up by a taxiing aircraft. Fortunately for this Whidbey Island Hellcat pilot, the propeller blades of the chewing aircraft stopped just short of the cockpit. 6 November 1943. (USN via Dave Lucabaugh)

(Below) White V88, an F4U-1 (02721) Corsair of VMF-313 has bellied-in at MCAS Mojave during transition training to the big Vought fighter. 27 December 1943. (via Dave Lucabaugh)

(Above) TBF-1 (06258) Avenger of VT-24 makes a 'textbook' ditching alongside USS BELLEAU WOOD. All three crewmembers can be seen exiting the sinking bomber. 2 September 1943. (USN-National Archives)

(Below) F6F-3 (65950) Hellcat of VF-3 went into the barrier while coming aboard USS HANCOCK. White 12 suffered only minor damage and went on to fight again. 21 November 1943. (USN via Dave Lucabaugh)

1944

(Above) FG-1A (14527) Corsair of VMF-115 was hit, along with several other parked aircraft, when a shot down Japanese bomber crashed into them. The ensuing fire burnt off most of the Glossy Sea Blue paint and fabric-covered surfaces. Leyte, P.I. 6 December 1944. (USMC)

(Below) SB2C-1 (00222) Helldiver of VB-15 was chewed up by a squadron mate that came through the barrier aboard USS HORNET. 7 January 1944. (USN via Dave Lucabaugh)

(Above) TBM-1 Avenger, White 8, of VT-51 takes the barrier while landing aboard USS SAN JACINTO. The pilot can be seen backing out of the cockpit as the engine pours smoke. 9 February 1944. (USN via Dave Lucabaugh)

(Below) PB4Y-1 (31939) of VPB-108 slid to a safe stop after making a forced landing with a hung up port landing gear at O'Hara Field, Apamama Island in the Gilberts. 9 January 1944. (USN-National Archives)

(Above) TBM-1, a General Motors/Eastern division-built Avenger, of VT-25 aboard USS COWPENS. During a raid on Truk, this plane was shot up by anti-aircraft fire that sheered the starboard horizontal stabilizer. In spite of this 'modification', the pilot completed an emergency landing aboard USS HORNET. 30 April 1944. (USN via Dave Lucabaugh)

(Below) FM-2 (16262) Wildcat of VC-10 belly landed aboard USS GAMBIER BAY. Damaged-beyond-repair at sea, the Wildcat was unceremoniously dumped off the side of the flightdeck. 18 June 1944. (USN via Dave Lucabaugh)

(Above) This OS2U-3 Kingfisher was damaged on landing and partially submerged. Crewmembers of the USS TANGIER, an aircraft recovery ship, worked to recover the damaged float plane. 16 April 1944. (USN via Dave Lucabaugh)

(Below) TBF-1D (24295) Avenger of VC-9 sunk too low during landing and hit the flightdeck ramp of USS SOLOMONS in a firey crash. Two of the three aircrew survived and were picked up by USS EISNER, the plane guard Destroyer. 25 March 1944 (USN-National Archives)

(Above) A Marine R4D-5 of VMJ-953 made a safe belly landing off the side of the runway at NAS Miramar, CA. The marking J55 was carried in White just forward of the fuselage National insignia. April 1944. (Clay Jansson)

(Below) Black 56, a PBM-3S Martin Mariner flying boat, tore the wing tip floats off and settled on the shallow bottom of Florida's Banana River. All the crew escaped uninjured. 10 May 1944. (USN via Dave Lucabaugh)

(Below) F6F-3 Hellcat, White 44, was attached to the Advanced Training Unit at Vero Beach, FL. It is pulled from the shallow waters off Ft Pierce where it was ditched after engine failure. 16 May 44. (USN-National Archives)

(Below) This PV-1 (34846) Ventura of HEDRON-1, FAW-5 skidded off the runway at MCAS Beaufort, SC after the starboard landing gear collapsed. 10 May 1944. (USN via Dave Lucabaugh)

(Above) A war-weary F4U-1 along with a pair of F4U-1As are unloaded from USS NASSAU off Muga Beach, CA. The two F4U-1As were last flown by VF-17. Aircraft such as these were frequently overhauled and assigned to training roles. 24 Jul 1944 (USN-National Archives)

(Left) This F6F-5 (58613) Hellcat, White 4 of VF-80, lost its belly tank while landing aboard the USS Ticonderoga. The resulting fire consumed much of the Hellcats port side. 1 July 1944. (USN via Dave Lucabaugh)

(Below) With battle-damage to its starboard wing this TBF-1C (48102), White 81 of VT-2 landed aboard USS HORNET. The eleven bomb marks just forward of the windscreen are mission strike markers. 13 June 1944. (USN via Dave Lucabaugh)

(Above) TBM-1C (25653) AVENGER of VC-11 carried the marking 3C on the tail in White with the 3 being repeated in Black on the cowling ring. Rocket launch rails are mounted beneath both wings. This crash landing by LT E P Goodin was the first on the newly constructed airfield at Falalop Island. 29 October 1944. (USMC)

(Below) TBM-1C (46203) Avenger of VC-21 crashed into the sea from the deck of USS MARCUS ISLAND. All three crewmen escaped uninjured. August 1944. (USN-National Archives)

(Above) This TBF-1 (24087) Avenger of CASU-24 belly landed just off the runway at NAS Wildwood, NJ. The starboard wing has partially folded. 11 Aug 1944. (USN via Dave Lucabaugh)

(Below) HE-1 (30277), a Piper-built flying ambulance version of the NE-1 Grasshopper, the military version of the J-3 Piper Cub. This all Gray HE-1 crashed at NAS Datona Beach, Fl. 18 September 1944. (USN via Dave Lucabaugh)

(Below) PV-1 (33280) Ventura of VB-130 slid off the side of the runway while making a crash landing at NAS South Weymouth, MA. 14 September 1944. (USN via Dave Lucabaugh)

(Above) This PB4Y-1 of Fleet Air Wing-7 is marked B-15. The Navy Liberator carries the Atlantic patrol paint scheme of Dark Gray over White. England, 27 December 1944. (USN via Dave Lucabaugh)

(Below) This SB2C-3 Helldiver of VB-15, with its starboard horizontal stabilizer and elevator shot up, has just recovered aboard USS ESSEX. 22 October 1944. (USN-National Archives)

(Above) This R4D-5 Skytrain, attached to Marine Air Group-45, has bellied into the surf off Falalop Island. All crew members escaped without injury. 3 November 1944. (USMC)

(Below) F6F-5 Hellcat of VF-4 scatters its load of rockets during a hard landing aboard USS ESSEX. Damage was minimal with none of the rockets exploding. 19 December 1944. (USN-National Archives)

(Below) F4U-1D of VBF-50 hits the barrier and noses over while landing aboard USS BATAAN. The checkerboard nose and tail colors are unknown. 9 December 1944 (USN-National Archives)

(Above) F6F-3 of VF-16 bellied into a cornfield near NAS Wildwood, NJ. This Hellcat, marked with a Yellow 9, carries the standard three tone finish of non-specular Sea Blue, Intermediate Blue, over White under surfaces. Fighting 16 first saw combat flying from USS LEXINGTON. 30 October 1944. (USN via Dave Lucabaugh)

(Right) TBM-1C (46253) of VC-12 crashed through the barrier while coming aboard USS CHARGER, and went over the port side still entangled in the barrier cables. Carrying a Black 9 on the tail this Avenger is painted in the Atlantic Gray over White scheme. 15 November 1944. (USN-National Archives)

This SBD-6 (54755) Dauntless of CASU-5 crashed on landing at NAAS Brown Field, CA. The engine has torn loose from its mountings and burned. The wrinkled fuselage carries the code C44 in White just forward of the National Insignia. 29 October 1944. (USN via Dave Lucabaugh)

(Above) *BLUE BARON*, an FG-1A (14143) Corsair of VMF-122 belly landed on the airfield on Peleliu Island. This Marine Fighting Squadron used a White geometric marking on the tail of its Corsairs. During World War II VMF-122 was credited with thirty-five kills. October 1944 (USMC)

(Left) F4U-1A Corsair of VF-10 (The Grim Reapers) flipped over on its back while landing aboard USS CORE. The deck crew quickly freed the trapped pilot. 16 December 1944. (USN via Dave Lucabaugh)

USMC R4D-5 Skytrain had an inflight fire in its port engine. Hastily landing at nearby NAS Miramar, CA, the transport's number one engine fell off as the plane came to a halt. Good fortune was with the crew that day. July 1944. (Clay Jansson)

1945

(Above) This JRC Cessna Bobcat slid to a stop off the end of the runway after making a wheels-up landing at NAS Atlantic City, NJ. The wood planking seen under the port nacelle is from the runway threshold fence. The aircraft carries an overall Silver paint scheme. 1945. (USN via Bob Lawson/The HOOK)

(Above) Returning from a raid on Chichi Jima, this TBM-1C (73321) of VT-86 crash landed aboard USS WASP. This Avenger, Black 313, carried the names *Baxter* below the cockpit, *Barry* on the nose and the Squadron name The Unlimited on the forward part of the fuselage. 18 February 1945. (USN-National Archives)

(Below) Marine FG-1A Corsair bellied in at Orote Airfield on Guam. The belly tank is smashed beneath the plane, and with the amount of fuel spilled on the runway, it is remarkable that there was no fire. January 1945. (Clay Jansson)

(Above) This JM-1 (66744) Marauder of VMJ-2 was destroyed while landing at its Saipan, Marshall Islands base. The Squadron insignia was carried on the nose, and was used as a photo aircraft in an Olive Drab over Gray paint scheme. July 1945. (Clay Jansson)

(Below) PB4Y-2 (59402) of VPB-118 was damaged while making a two wheel landing. The starboard main gear failed to extend when the Privateer was coming in at Motoyama Airfield No 1 on Iwo Jima. March 1945. (USMC)

(Below) With damage to both wings, this TBM-3 of VT-17 recovered aboard USS BUNKER HILL. This Avenger carrying the three tone paint scheme has the colorful Yellow nose band used by Air Group-17. 19 February 1945. (USN via Dave Lucabaugh)

(Below) A combat damaged TBM-3 Avenger of VC-94 carrying the markings of USS SHAMROCK BAY crashed while coming aboard USS PETROF BAY. The Avenger broke in half just aft of the turret gun. 29 January 1945. (USN via Dave Lucabaugh)

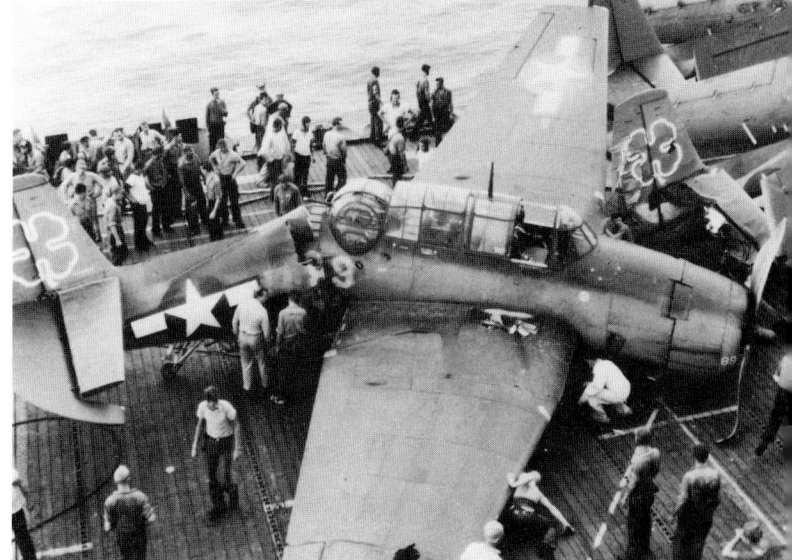

(Above) This Curtiss XBTC-2 (31401) suffered propeller and wing damage when this prototype torpedo bomber's port landing gear collapsed. The Pratt & Whitney R-4360 powered aircraft never reached production status. 3 Mar 1945. (Curtiss Wright via Hal Andrews)

(Below) TBM-3 Avenger of VT-6 off USS HANCOCK carries a replacement rudder that retains its original markings, not an uncommon sight during wartime. When time permitted these replacement parts were repainted. 27 Mar 1945. (USN-National Archives)

(Left) SC-1 (35490) Seahawk of VN-7D8, a training squadron, was heavily damaged when it nosed over during a landing on Pensacola Bay. This Seahawk never flew again. 23 March 1945 (U.S. Navy)

(Below Left) F4U-1D Corsair of VMF-512 was dropped from the loading crane while being hoisted aboard USS GILBERT ISLANDS at San Diego. The EE-64 markings were in Yellow. 11 March 1945. (USN-National Archives)

(Below) SB2C-4 (63031) Helldiver of VB-93 pushed its luck to the maximum, having slid into the high explosives magazine at Otis Field, Camp Edwards, MA. Fortunately, there was no fire or explosion. 6 April 1945. (USN via Larry Webster)

(Above) F7F-3P (80424) Tigercat of VMD-254 is carrying a D13 in White on the tail. This photoplane slid to a stop and burned just off the runway at Kadena Field, Okinawa. 1945. (Clay Janson)

(Below) This PB4Y-1 of VD-5 with a huge White 52 on its tail bit the dust on the taxiway when the port main landing gear collapsed. A heavily weathered three tone camouflage finish is carried on this Liberator. Philippines, 10 March 1945. (USN via Dave Lucabaugh)

(Right) FM-2 (55219) Wildcat of the Carrier Qualification Training Unit from NAS Glenview, IL fouled the deck when it hit the barrier on the training carrier USS SABLE. Carrying the markings M-3 on its fuselage, this tall-tail Wildcat suffered little damage. After repairs, it flew again. May 1945. (USN via Dave Lucabaugh)

(Below) This high-time, war-weary USMC SBD-5 carrying a White 713, has the name QUEENIE just forward of the 114 mission markers stenciled under the cockpit. Valuable equipment was removed with the remainder of the Dauntless being considered scrap. Malaban, Mindaneo, Philippines, 24 July 1945. (USMC)

(Above) Grumman J2F-5 Duck amphibian marked H23 wore an overall Silver finish. Powered by a Wright Cyclone R-1820-54 900 hp engine, it went off the runway and flipped on a Guam airfield. September 1945. (Clay Jansson)

(Below) TBM-3 (68629) Avenger of VT-29 goes over the port catwalk during its takeoff from USS CABOT. Remarkably, the pilot HH Skidmore successfully flew it away with minor damage. 1 April 1945. (USN-National Archives)

(Above) White 70, an F6F-3 Hellcat, missed all the wires and barriers while landing on USS SOLOMONS and slammed into another Hellcat causing heavy damage to both aircraft. 3 May 1945. (USN-National Archives)

(Below) F4U-1A (17801) Corsair of VBF-3 made a wheels up landing at NAS Wildwood, NJ. This Corsair carried the White stincelled markings G2 stacked on the tail. 19 June 1945. (USN-National Archives)

(Left) F6F-5 (70803) Hellcat of VF-22 momentarily balances on its propeller after tangling with the barrier aboard the escort carrier USS TAKANIS BAY. The pilot is bracing himself in the cockpit for the coming jolt. The Grumman Hellcat was a sturdy airframe that could take a great deal of punishment, and this one survived to fight another day. 30 Jul 1945. (USN via Dave Lucabaugh)

(Below) TBM-3 Avenger of VT-83 was heavily damaged by Japanese anti-aircraft fire while on a bomb strike over Okinawa. The Avenger struggled back to USS ESSEX where it made a safe splash down near the plane guard Destroyer. All the crew was rescued. 1 April 1945. (USN-National Archives)

(Above) F6F-5 (79376) Hellcat of VBF-11 took the barrier and nosed over aboard USS TAKANIS BAY. This Hellcat carried the S73 marking in White on the fuselage and top side of the starboard wing. Right through the closing days of the war, the Navy carried on a voracious flight training program. 11 August 1945. (USN via Dave Lucabaugh)

(Above) This SNJ-4 (26786) of OTU T-4 carried the markings JON 077, and a Green band just aft of the fuselage National insignia, and on the wings. While taxiing, the starboard landing gear collapsed causing little damage other than a scrapped wingtip, flap, and aileron. The look on the Captain's face says it all! 13 August 1945. (USN-National Archives)

(Below) This FM-2 Wildcat of VC-96 took the barrier after its arresting hook failed to engage the wires while coming aboard USS RUDYARD BAY. The White geometric marking on the tail was carried on both sides. The White drop tanks were a carry over from the three tone paint scheme. 1 Apr 1945. (USN-National Archives)

(Below Right) TBM-3 (22910) Avenger of VT-40 carries the White D 92 markings on the fuselage just forward of the National insignia. This three tone Avenger incurred damage to the propeller, both wingtips and broke the tailwheel. USS SUWANEE, 28 June 1945 (USN-National Archives)

SB2C-1 Helldiver of VB-6 broke a tailwheel while landing aboard USS YORKTOWN during Carrier Qualification Trials. The deck crew is poised to rush to the aircrew's assistance. May 1943 (USN-National Archives)

F6F-3 Hellcat of VF-25 broke the starboard landing gear during a hard landing on USS COWPENS. The Hellcat was returning from a strike on the Marshall Islands. November 1943 (USN-National Archives)

F4U-1A (17884) Corsair of VMF-214, Boyington's Blacksheep Squadron, sits in Torokina's boneyard alongside a USAAF P-39. This Corsair was demolished when it flipped over during an attempted belly landing. January 1944 (National Archives)

This PBM-3 Mariner of VP-208 crash landed in the Atlantic off the West Indies. 5 May 1943

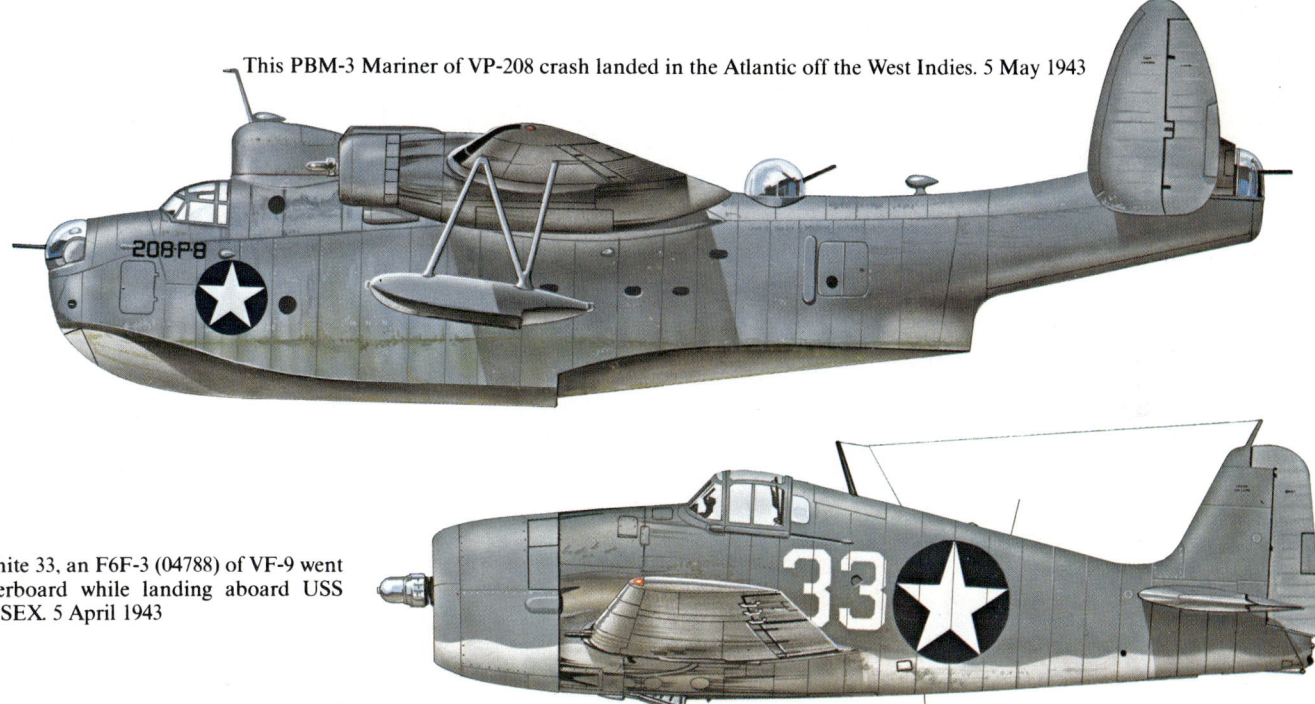

White 33, an F6F-3 (04788) of VF-9 went overboard while landing aboard USS ESSEX. 5 April 1943

1946

(Above and Below) F4U-4 (81437) Corsair of VF-4 lost its belly tank during landing aboard USS TARAWA. The ruptured tank caused an intense fire which burned away all the fabric and badly scorched the metal surfaces. 7 February 1946. (USN-National Archives)

(Above) F6F Hellcat of VF-97 nosed over during roll-out. The cause of this particular accident is not known, but nose overs usually occurred as a result of hard braking and was not considered an unusual event at an Air Station. 1946. (U.S. Navy)

(Above Right) An SNJ-5 drags its wingtip during a hurricane fly-away at Bronson Field, Jackson, MS. The Texan was notorious for its sensitive landing characteristics. People who have flown it swear that there are two kinds of pilots — those who *HAVE* groundlooped it, and those who *WILL*! 1946. (Authors Collection)

(Below Right) FG-1D Corsair assigned to an Advanced Training Unit flipped on to its back while landing at an Auxiliary Air Station near NAS Jacksonville, FL. The pilot escaped serious injury. 1946. (U.S. Navy)

This war-weary PB4Y-2 (59522) Privateer of VPB-109 has been stripped of its combat armament. *Poison Ivey* was stencilled in White just beneath the turret gun position. Camp Kearny, CA. 1945 (Lou Darden)

SBD-5 (09752) Dauntless of Marine Air Wing-3 crashed and broke in half during a training flight at MCAS Cherry Point, NC. 20 August 1943

Coded Black 24, this TBF-1 Avenger was lost when it went over the side of USS CARD when it crashed while landing. 26 December 1943

PB2Y-3 *Betty B* coded P-70 in Black. This Coronado Patrol Bomber was beached at Salinas, Brazil. 13 October 1943

This F4U-4B (63059) of VMF-312 belly landed in the Korean sand. The pilot wears a Blue and White checkerboard on his helmet matching the Corsair's nose band. 1952 (Authors Collection)

This TBF-1 (06258) Avenger of VT-24 was ditched and lost alongside USS BELLEAU WOOD. 2 September 1943

White 3 an F4U-1A of VF-17 flown by LT (jg) FJ (Jim) Streig carries six kill marks. Piva Field, Bougainville. February 1944

F6F-3 of the Advanced Training Unit based at Vero Beach, FL. This Hellcat was ditched in shallow waters. 16 May 1944

(Above) F6F-3 (42167) Hellcat of OTU, VTB-2 going over the port side of USS SOLOMONS, ending the last flight of this Hellcat. The pilot's job at this point was to keep the nose up and wings level for the impact of the splash down. The fuselage marking MF was carried in White. 15 January 1946. (USN via Dave Lucabaugh)

(Right) This F4U-1A was found underwater off Oahu, Hawaii during the early 1970s by skindivers. The Corsair was apparently successfully ditched as evidenced by the open position of the canopy. Tropical Angelfish which abound in these warm waters swim around the hulk. To have been submerged in salt water for such a period of time, the F4U appears to be in relatively good condition. The date of the crash was estimated to have taken place during 1946. (Joe Henderson)

(Below and Below Right) F4U-4 (81792) of VF-75 went over the side while coming aboard USS FRANKLIN D ROOSEVELT. This Corsair sank rapidly but not before the pilot was able to escape. VF-75 briefly carried a White band across the tail containing VF-75 or just VF in black. 26 April 1946. (USN-National Archives)

1947

(Above) F6F-5 (77469), based at NAS Anacostia, bellied in at Wilmington, NC while shooting touch-and-go landings. This Hellcat carried its station name in large White letters beneath the port wing and in smaller letters across both sides of the tail. 1947. (Paul J. McDaniel)

(Left) F8F-1 (95395) Bearcat of VF-19A carrying the White code B 122 on the tail and fuselage. While coming aboard USS TARAWA this Grumman Bearcat missed the arresting wires and floated into the barrier. Seen on its second bounce, it has already lost its starboard landing gear and wingtip. The 'B' on the fin was Yellow, and the rudder trim tab stripes were Yellow and White. 5 August 1947. (Bob Lawson)

(Below) During 1947 this USMC F4U-4B crashed in a clearing among the pine trees in the woods surrounding Camp LeJeune, NC. The Corsair was stationed at MCAS Cherry Point and is being recovered and trucked out with the cooperation of the USMC by a civilian group wanting to restore the machine. 31 September 1967. (USMC)

F4U-5N (124453) of VC-3, **Annie Mo**, a Corsair night fighter flown by Navy Ace, LT Guy P. Bordelon. The crash occurred at the hands of an Air Force pilot. Korea, 1953 (W.F. Gemeinhardt, M/SGT USMC, Retired)

BLUE BARON, an FG-1A (14143) Corsair of VMF-122, the belly landed at Peleliu Island. October, 1944

This SNJ-4 (26786) was damaged when the starboard landing gear collapsed. 13 August 1945

PB4Y-2 (59402) of VPB-118 sustained strike damage when landing at Motoyama Airfield #1 on Iwo Jima. March 1945

F4U-4 (81710) of MAMRON-15 (Marine Aircraft Maintenance Squadron) nosed-over while landing, chewing up all four tips of the 13 foot Hamilton Standard propeller. Korea, 1953 (H.R. Rued via Ron Picciani Aircraft Slides)

White 713 a War-Weary Marine SBD-5 Dauntless carries the name *QUEENIE* beneath the cockpit just forward of 114 mission markers. Malaban, Mindanao, P.I. 24 July 1945

White 52, a PB4Y-1 of VD-5 ends its career in a drainage ditch on its Philippine Island base. 10 March 1945

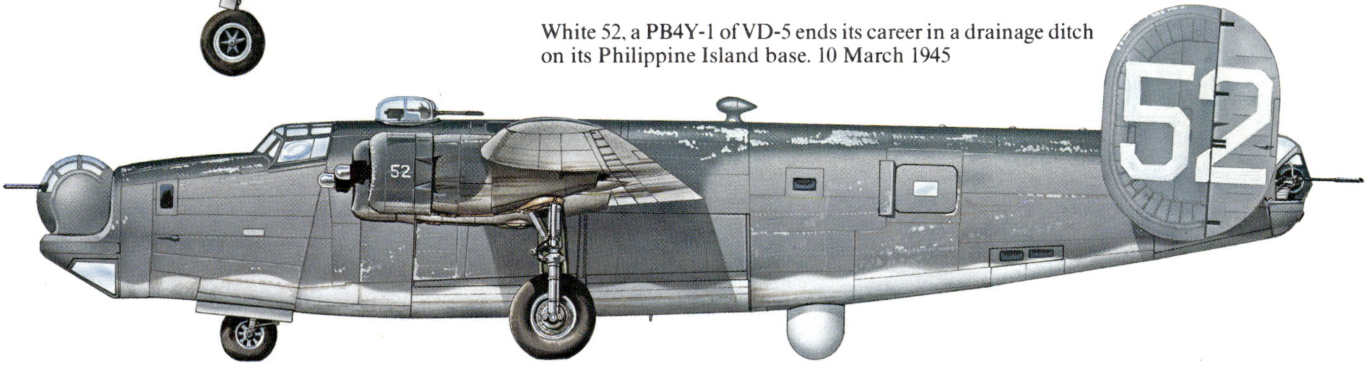

SB2C-4 (63031) Helldiver of VB-93 crashed into a high explosive magazine at Otis Field, MA. 6 April 1945

(Above and Below) Flown by Ensign R T Smith, this F4U-4 (82051) of VF-17A (EX-VF-82), started spraying engine oil all over the cowling and windscreen completely blocking out forward visibility. Keeping his professional 'cool', Smith landed on USS RANDOLPH at an angle in order to see the flightdeck by looking out of the open the canopy. On landing, the starboard main gear and the propeller dug into the deck, but Smith walked away and the Corsair flew again. The Corsair's propeller hub and diamond-shaped marking on the tail are Red. Mediterranean Sea, 23 February 1947. (R.T. Smith via Bob Lawson/The HOOK)

1948

This N3N-3 hit the water hard enough to break off the front of the main pontoon and flip the Boeing Kaydet trainer onto its back. The U.S. Navy used N2N and N3N biplanes to train thousands of aviators. 1948. (U.S. Navy)

(Below and Below Left) This AD-4Q (124056) Skyraider in the markings of Carrier Air Wing-17 off USS FRANKLIN D ROOSEVELT, blew a main engine seal while inflight. CDR W.N.Leonard landed his oil-covered Skyraider at the nearest field, Wilmington, NC. Leonard is all smiles after climbing out of the cockpit. 1948. (Paul J. McDaniel)

An AD-4N (127003) of VA-95 awaits repair along with a Panther of VF-151 and a Skyraider of VF-194. Korea, 1953 (via J.E. Michaels-JEM Aircraft Slides)

Yellow 9, an FM-2 Wildcat of VC-96 crashed while coming aboard USS RUDYARD BAY. 1 April 1945

F4U-4 (82051) Corsair of VF-17A crash landed aboard USS RANDOLPH. 23 February 1947

F7F-3P (80424) Tigercat of VMD-254 crashed and burned at Kadena, Okinawa. December 1945

AD-3 (122751) Skyraider of VMA-121 belly landed at airfield K-47 in Korea during the Winter of 1952. (Ron Picciani Aircraft Slides)

AD-4Q (124056) Skyraider of Carrier Air Group-17 from USS FRANKLIN D ROOSEVELT. *CDR.W.N. LEONARD* is stenciled beneath the cockpit. 1949

F9F-2 (123494) Panther of VF-21 tore through the barriers aboard USS MIDWAY and into a fiery collision that ended up in the sea. 11 November 1951

F2H-2 (124997) Banshee of VF-12 crashed into the barrier aboard USS WASP. 22 September 1952

(Above) F6F-5 Hellcat of ZX-11 carried the unusual tail model XL. The Hellcat was attached to Airship Development Squadron-11, a lighter than air unit. This aircraft was generally used by LTA pilots to keep their flying proficiency in both classes. This crash ended the flying days of White 21. 1948. (Authors Collection)

(Right) F4U-4 (81795) Corsair of VMT-2 lost its starboard landing gear and hit the barrier while recovering aboard USS MONTEREY. White 56 came to a rest just short of the ship's island. 1948. (Clay Jansson)

(Below) PB4Y-2 (569803), a weary patrol bomber, is parked at the Naval Air Storage Facility at Litchfield Park, AZ. This Privateer last served the U.S. Navy with VPHL-8. 1948. (via Alan Blue)

This F4U-4B (62974) Corsair of VMF-312 caught the arresting wire, took a high bounce, and slammed back to the flightdeck with such force that the port landing gear collapsed. The impact bent the propeller and nailed the Glossy Sea Blue Corsair to the deck. The pilot was thoroughly shaken but had no serious injury. August 1948. (Clay Jansson)

(Above and Below) This F4U-4 of VF-74 was unable to lower the main landing gear and had to come aboard USS Bon Homme Richard wheels up. Chewing up the flightdeck, the belly landing was made after returning from a mission over Korea. September 1952 (Joe Michaels-JEM Aircraft Slides)

1949

(Above) This SNJ-5 Texan flipped over after dragging the port wing and rests on its back on a snow covered runway at NAS Glenview, IL. Black lettering was applied over the natural Aluminum skin. 1949. (Clay Jansson)

(Below) F6F-5 (80199) from the Naval Air Reserve Training Unit at NAS New Orleans was damaged during a wheels-up landing during on a hop to the Pensacola Naval Air Station. This Hellcat carried its home station insignia just under the windscreen. 1949. (Bill Crimmins)

This F4U-4 (81407) Corsair of VF-74 lost power on the final approach to USS PHILIPPINE SEA. Realizing he was unable to make the flight deck, the pilot successfully put the Corsair down on the water and was quickly rescued. 3 March 1949. (USN-National Archives)

(Above) F4U-4P (63050) of VMP-254, a USMC photo Corsair carrying the modex WT, crashed while landing aboard USS HAIROKO. Determined to be beyond economical repair, the F4U was given the deep-six. This Corsair was the final F4U-4P built out of a total of nine produced by Chance Vought. February 1949. (Clay Jansson)

(Right) TBM-3U (85959) of VU-7 nosed over while landing. With Yellow and Red trimmed flying surfaces, this colorful Avenger was used by Utility Squadron-7 (VU-7) for target towing. A number of TBMs survived WW II and served during and after the Korean War. 9 May 1949. (Capt. Floyd Harris via Bob Lawson, The HOOK)

(Above) F8F-1 (94929) of VF-11 is marked with a White 'T' on the tail and the numerals 109 in White on the fuselage. This Bearcat made a successful night belly landing at NAS Cecil Field, FL. 26 July 1949. (Capt. W.R. Stuyvesant via Bob Lawson/the HOOK)

(Below Left, Below Center, and Below) This F4U-4 (96771) of VF-74 *almost* took a wave off from the LSO while recovering aboard USS PHILIPPINE SEA. Applying full throttle caused a fatal torque roll that put the Corsair on its back and into the sea. 1 January 1949. (USN-National Archives)

(Below and Below Right) F4U-5 of VF-21 marked White M 117 had the belly tank catch fire after coming aboard USS CORAL SEA. The pilot safely exited the flaming Corsair but the intense fire consumed the F4U well past the point of possible salvage, and the charred hulk was pushed off the number two elevator. 15 June 49. (USN-National Archives)

(Above) After a wheels up landing in the countryside near Wilmington, NC, this TBM-3E (91344) Avenger of VU-1 was recovered by a Navy salvage team, loaded on a flat-bed trailer and trucked back to NAS Norfolk for repair. This Avenger was easily repaired and flew again. 1949. (Paul McDaniel)

(Below) This FG-1D (88446) Corsair from the Naval Air Reserve Training Unit at NAS Oakland took a bad bounce while doing Field Carrier Landing Practice (FCLP) at NAS Alameda, CA. Unable to recover the bounce in time, the Corsair pilot rode it in for a grinding crash on the duty runway that broke it completely in half. 6 February 1949. (Larry Smalley)

1950

(Right) An F4U-4B from Air Group-11 began pouring smoke from the engine right after takeoff. With limited vision, the pilot put the Corsair into a steep dive, both to get down as quickly as possible an attempt to put out the fire. A hasty but successful water landing was made. 22 October 1950. (USN-National Archives)

(Below) AD-3Q (122875) Skyraider of VC-33 (Nighthawks) encountered heavy anti-aircraft fire while on a mission over Pyongyang, North Korea. Riddled with holes in the tail section, Ensign R.H. Rohr carefully turned back South, heading for the nearest divert field. The Skyraider made it back safely. All lettering on this Glossy Sea Blue AD was White, including the SS modex. 24 November 1950. (USN-National Archives)

(Above) This FG-1D photo-Corsair suffered a failure of the starboard landing gear causing the fighter to leave the paved runway shear off the other landing gear and slide to a stop. The pilot exited the cockpit uninjured. The aircraft was assigned to the Naval Air Training Unit at South Weymouth, MA. 1950. (Authors Collection)

(Right) AU-1 (133837) of VMF-225 carries the USMC modex 'WI' in White on the tail. During a landing at NAS Niagra Falls, NY, the plane veered off the runway into the grass and to destruction. This Corsair was the seventh from the last one built. 1950. (Mort Hartman)

(Below) F9F-5P (126268) Panther of VC-61 comes aboard USS BOXER off Korea. As the weight of the Panther came to bear on the nosewheel, it collapsed sending the damaged plane skidding down the flight deck. The pilot escaped injury. Most Navy planes of this period wore Glossy Sea Blue paint, a handful were left in anodized natural Aluminum to test this finish against corrosion. 1950. (USN via Don Spering, A.I.R.)

1951

(Above) This AD-3W (White 84) of VC-11 sustained damage from anti-aircraft fire while on a mission up North. Escorted by another AD-3W (White 39), the damaged Skyraider headed for a divert field rather than going back to USS BOXER. White 84 was quickly repaired and returned to the fighting. Korea 13 June 1951. (USN-National Archives)

(Below) This AD-4 (123933) Skyraider of VA-195, on a bridge-busting mission South of Wonson, Korea took an anti-aircraft hit blowing away a substantial portion of its vertical stabilizer. LT Phillips successfully brought his crippled plane back to USS PRINCETON. The scraped paint on the underside of the wing was due to the installation and firing of 5 inch HVAR rockets. 30 April 1951. (USN-National Archives)

An F9F-2 (123494) Panther of VF-21 recovering aboard USS MIDWAY failed to pick up a wire. Tearing completely through the barriers, the Panther slammed into F9Fs spotted on the forward deck. The fiery collision resulted in the loss of three jets, all of which ended up in the sea. All markings on the Glossy Sea Blue Panthers are White including the bands around the wings and fuselage of number 115 and 116 in November 1951. (USN-National Archives)

(Above) F4U-4 of VF-713, a Denver Reserve outfit called to active duty during the Korean War, carries the modex 'H' in White on the rudder and upper starboard wing. Coming aboard USS ANTIETAM, this Corsair had a belly tank fire which was quickly extinguished, causing minimal damage. 24 November 1951. (USN-National Archives)

(Left) AD-2 (122310) Skyraider of VA-702 flying off USS BOXER was hit by anti-aircraft fire during a bomb strike in Korea. LT (jg) W. Sullivan elected to land *Jinx* at a nearby divert field without further damage. The damaged rudder and shredded sheet metal was replaced and the plane was flown back to the carrier. 13 June 1951. (USN-National Archives)

(Below) This AD-4L (123968) Skyraider of VA-728 crashed into the barrier while coming aboard USS ANTIETAM. The engine broke away on impact resulting in a flightdeck fire. The deck crew quickly extinguished the fire, but the Skyraider was beyond use and was pushed off the flightdeck. 10 July 1951. (USN-National Archives)

(Below Left) F4U-4 of VF-653 crashed while coming aboard USS PHILIPPINE SEA and went off the port side of the deck where it wrapped itself around a 5 inch gun tub. This Corsair, marked H-320 in White, was assigned to the Reserve unit called to Active Duty from NAS Akron, OH. December 1951. (Danny Tyree via Jim Wiedie)

(Above) F7F-3P (80431) Tigercat of Marine Aircraft Maintenance Squadron 33 (MAMRON-33) went off the embankment of a Korean Airfield entangling itself in telephone wires and burning. The pilot was fatally injured and the Tigercat was reduced to scrap metal. 22 February 1951. (Clay Jansson)

(Left) Navy pilot, LT Harold Reutebuch displays a big grin of relief at his good fortune. While flying an AD-4 Skyraider with VA-923, a Reserve outfit from NAS St Louis, MO the Skyraider took an anti-aircraft hit in the canopy just inches behind Reutebuch's head. USS BON HOMME RICHARD, 8 July 1951. (USN via Harold Reutebuch)

(Left) F9F-2 (123625) Panther of VF-781 loses the forward section of fuselage while landing aboard USS BON HOMME RICHARD. This Panther carried D 117 in White and was returning from a strike over Korea. With the replacement of the forward fuselage, the Panther was again airworthy. June 1951. (USN-National Archives)

(Above) F8F-2 Bearcat carrying the markings VW-402 was assigned to the Officer C.I.C. School at NAS Glenview, IL. The tall tailed Bearcat nosed over and caught fire, burning the starboard flap and scorching the fuselage's metal bottom. 1951. (via Clay Jansson)

(Right) AD-4N Skyraider of VC-35 was hit in multiple locations by North Korean anti-aircraft fire. Flying from the USS BON HOMME RICHARD as part of Task Force 77, she was not seriously damaged. June 1951. (USN-National Archives)

(Below) R5D-3 of VMR-152 crashed in the mountains while returning from a supply run to Korea. The modex 'WC' was carried in Black on the tail of this natural Aluminum Skymaster. There were no survivors. Japan, 30 May 1951. (Clay Jansson)

1952

(Above) AF-2W (123117) Grumman Guardian of VS-931 carrying the White modex 'SV' teeters on the bow of USS SICILY at dockside. The damaged plane was removed by crane, repaired, and returned to service. 15 April 1952 (via Bob Lawson/the HOOK)

(Left and Bottom Left) PB4Y-2 of VP-9 lost its port main gear while landing at NAS Barbers Point, Hawaii. The crash caused unrepairable damage to the Privateer. The modex 'CB' was carried in White on the tail. August 1952. (Authors Collection)

(Below) F4U-4 of VMF-312 returning from a combat mission over North Korea. When 1st LT Paul Manning lowered the gear, he discovered that the left wheel would not extend. Making a one wheel landing, Manning successfully put the Corsair down on the steel runway and slid to a safe stop with surprisingly little damage to the plane. November 1952. (USMC)

(Above and Right) AD-4 (128925) Skyraider of VA-65 landed aboard the deck of USS BON HOMME RICHARD. An F4U-4 (81087) was taxiing into launch position when the AD chewed its way up the F4U's fuselage. The Corsair immediately burst into flames and was consumed. 22 January 1952. (USN-National Archives)

(Below) F9F-2 Panther of VF-51 completed its mission over Korea and was heading back to base when anti-aircraft fire ruptured a fuel line. Jet fuel accumulated in the fuselage but did not burn until a spark set it off on landing turning the Panther's rear fuselage into a blazing torch. LT (jg) R.E. Rostine, the pilot, escaped safely. 2 January 1952. (USMC)

(Above) AD-2 (122234) Skyraider of VA-859 hit the barrier while coming aboard USS TARAWA. The nose-over crash caused an engine fire that quickly spread to burn the forward part of the Glossy Sea Blue fuselage and portions of both wings. This aircraft, carrying White markings E 506, when not at sea was stationed at NAS Jacksonville, FL. 7 July 1952. (USN-National Archives)

(Below) F9F-2 (127147) Panther of VF-837, a Reserve squadron from NAS New York was called to active duty serving as a part of Air Group-15 during the Korean War. While recovering aboard USS ANTIETAM after a strike in Korea, this Panther experienced a collapsed nosewheel causing only minor damage to the nose section. After repairs it flew again. February 1952. (USN via Hal Andrews)

(Above) This TBM-3 (85970) of VS-21 carries the modex 'BS'. While landing aboard USS BADOEING STRAIT, the Avenger left the flightdeck and crashed into the sea. The pilot can be seen on the starboard wing. At this point in time, very few of the WWII vintage Avengers remained in service. 9 April 1952. (USN-National Archives)

(Below) Assigned to OPERATION SKI JUMP II in Alaska, R4D-4 (12417) crashed during an attempted takeoff when the port main landing gear collapsed. This ski-equipped Skytrain was never flown again and was left derilict on the ice. The colorscheme was overall natural Aluminum with Red Artic markings on the rear fuselage, tail and tops of the outer wing panels. The missing propeller from the port engine can be seen behind the starboard wing. 27 January 1952. (USN-National Archives)

(Above) F2H-2 (124997) of VF-12, marked in White, T 208 is part of Air Group-1 participating in OPERATION MAINBRACE in the Arctic Ocean. This Banshee crashed into the barrier while landing aboard USS WASP. Damage was minimal to the nosewheel area. 22 September 1952. (Colonel Ed Mason via Bob Lawson/the HOOK)

(Right and Bottom Right) This F2H-2P (125686) of VMJ-1 was damaged by anti-aircraft fire over Korea. The pilot, Master Sergeant Meyers did a skillful job in getting his photo-Banshee back on the ground at K-3, Korea. March 1952. (via Clay Jansson)

(Below) Captain James McDaniel of VMF-312 points to where the bullet that shattered the windscreen of his F4U-4 entered. Captain McDaniel brought his damaged Corsair back to USS BATAAN off Korea. May 1952. (USN-National Archives)

(Above) F9F-2B of VF-24 was heavily damaged during a hanger deck fire and explosion aboard USS BOXER. The White modex 'M' was carried on the tail of the Panther. In the background is the burned remains of an F4U-5NL (124517) Corsair nightfighter of VC-3. 6 August 1952. (USN-National Archives)

(Below) F9F-5 (126019) of VF-837, a Naval Reserve unit called-up from NAS New York, floated up the flightdeck and into the barrier which was rigged aboard the USS PHILIPPINE SEA. This Grumman-built Panther Jet was a part of AIR GROUP-15 operating as a part of the all-Reserve force onboard the USS ANTIETAM. Korea, January 1952. (USN-National Archives)

(Above) F9F-2B (123507) of VMF-232, marked with a White WT 5, left the runway and climbed an embankment at MCAS Kanohoe Bay, Hawaii. Surprisingly, only minor damage was sustained. 1952. (via Clay Jansson)

(Below) Resting on 55 gallon drums, this F4U-5N of VMF(N)-513 was demolished in a crash during the Korean War. The FLYING NIGHTMARES were the only USMC squadron to use overall Matte Black with Red markings on their F4U-5N Corsair and F7F Tigercat nightfighters. Korea. 1952 (Authors Collection)

(Right) The pilot of this F4U-4B of VF-53 returned from a mission over Korea and found that the landing gear would not extend. He successfully made a wheels up landing aboard USS ESSEX, damaging the Corsairs prop and flaps and chewing up some flightdeck. 1952. (USN-National Archives)

1953

AJ-2 (130410) Savage of VC-6 *literally* lost its port engine during an arrested landing aboard USS ESSEX. The propeller blades crunched through the fuselage just forward of the pilots seat. The impact of landing also tore loose the port wingtip fuel tank. 10 November 1953. (W.E. Scarborough via Bob Lawson/the HOOK)

(Above and Left) This AD-4Q (124055) of VF-194 lost power on takeoff from USS BOXER. LT (jg) Joe Kagi held the wings level and made a water landing. On impact, the Skyraider had its engine and both outer wing panels torn off, but ended up floating high on the ocean. Kagi escaped unharmed. 16 July 1953. (USN-National Archives)

(Above) This F4U-4 Corsair of VF-152 floated over the arresting wires and barriers, tearing off the tail of another Corsair from the squadron. The crashing F4U-4 then went over the port side with the pilot making a safe water landing seconds later. This took place aboard USS PRINCETON just prior to its Korean combat tour. 12 February 1953. (USN-National Archives)

(Below) The tailhook failed to extend as this AD-6 (128959) of VA-65 approached USS YORKTOWN. The big Skyraider took the barrier and nosed over. As the propeller dug into the deck, one of the blades broke free. The deck crewmen are attaching lines to the aft end of the fuselage to pull it back down on the deck. December 1953. (USN via Hal Andrews)

(Above) F4U-5N Corsair nightfighter of VC-4 carried the codes NA 55 in White. The arresting hook caught a late wire as the propeller and landing gear became entangled in the barrier. The pitot tube on the port wing is the only apparent damage. 14 April 1953. (USN-National Archives)

(Right) During the closing days of the war, this F9F-2B of VF-91 on a combat mission over Korea took an anti-aircraft hit and lost its nose assembly. LT (jg) Batten managed to maintain control of the damaged Grumman Panther, flying it back to a safe arrested landing aboard USS PHILIPPINE SEA. 12 July 1953. (USN-National Archives)

(Below) This AD-4 of VA-728 marked H 510 in White, made a wheels up crash landing while coming aboard the USS ANTIETAM. The big Glossy Sea Blue Skyraider slid into the barrier and came to rest against the ship's island. This landing was especially hazardous since a bomb remained on the outboard station of the starboard wing. The pilot escaped injury. 25 February 1953. (USN-National Archives)

An SNJ-5C from a Carrier Qualification Training Unit (CQTU) attempted a landing aboard USS MONTEREY off the California Coast. The Texan got away from the pilot and rolled to the starboard, raking the flightdeck and breaking off several feet of wingtip. The SNJ continued over the deck, hitting the water right-side up. 28 August 1953. (Bob Lawson)

(Above) F9F-4 (125212) of VMF-115 plowed its way into a cultivated field just off the runway at K-3, Korea, when the Panthers landing gear refused to lower. 1st LT L.A. Merriman (formerly a major league baseball player) brought the Grumman jet in for its belly landing. 21 April 1953. (Clay Jansson Collection)

(Left) This F4U-5N (124453) of VC-3 experienced a devastating crash at the hands of an Air Force officer who wanted to fly the plane. Unfamiliar with the hefty torque of the big P&W R-2800-32W engine, the Corsair rolled over on the ground during the attempted takeoff. NP 21 was the aircraft of Navy Ace Guy P. Bordelon. Korea 1953. (Henry Covington)

(Below) AD-4B (132286) of VA-75 made a crash landing and burned at the end of runway 30 at NAF Charleston, RI. The Skyraider suffered strike damage as a result of the crash. Scrambling clear of the burning plane, the pilot escaped serious injury. 7 January 1953. (USN via Larry Webster)

(Above) F2H-2 of VF-11 was marked T 101 in White and was part of Air Group-101 aboard USS KEARSARGE. This Banshee participated in raids over Korea during the closing days of the war. The crash occurred aboard USS VALLEY FORGE. 1953. (Clay Jansson)

(Below) F8F-1 (95211) from the Naval Air Training Unit at St Louis, Mo made a perfect centerline landing. Unfortunately the wheels were up, and the Bearcat bellied down the runway amid a shower of sparks. There was no injury, and with a replacement propeller, an engine check, and some sheet metal work, the F8F was quickly in the air again. Goodfellow AFB, TX. 15 July 1953. (USN via Roger Besecker)